Nelson Orlando Álvarez Nicoliello

EARTH LOBRIZE Common Annelid

Nelson Orlando Álvarez Nicoliello

EARTH LOBRIZE Common Annelid

Use of earthworm (lumbricus terrestris l.) for humus production

ScienciaScripts

Imprint

Any brand names and product names mentioned in this book are subject to trademark, brand or patent protection and are trademarks or registered trademarks of their respective holders. The use of brand names, product names, common names, trade names, product descriptions etc. even without a particular marking in this work is in no way to be construed to mean that such names may be regarded as unrestricted in respect of trademark and brand protection legislation and could thus be used by anyone.

Cover image: www.ingimage.com

This book is a translation from the original published under ISBN 978-613-9-43828-0.

Publisher:
Sciencia Scripts
is a trademark of
Dodo Books Indian Ocean Ltd. and OmniScriptum S.R.L publishing group

120 High Road, East Finchley, London, N2 9ED, United Kingdom
Str. Armeneasca 28/1, office 1, Chisinau MD-2012, Republic of Moldova, Europe
Printed at: see last page
ISBN: 978-620-8-14299-5

Copyright © Nelson Orlando Álvarez Nicoliello
Copyright © 2024 Dodo Books Indian Ocean Ltd. and OmniScriptum S.R.L publishing group

INDEX

1. INTRODUCTION

Agroecology, as the science of ecology applied to agriculture, is uniquely distinguished by its recognition of social and ecological assessment and the inseparability of social and ecological systems. In Venezuela, since 1999, the development of this branch has been boosted by the approval of the new national constitution, which includes the protection of nature and in its article 305, the following text: The State shall promote and develop agroecology as a strategic basis for integral rural development, in order to guarantee the food security and sovereignty of the population, understood as the sufficient and stable availability of food at the national level and timely and permanent access to it by the consuming public.The use of chemical fertilisers as suppliers of agricultural nutrients has been considerably reduced, due to the high price they have on the world market and also because of the international trend to reduce chemicalisation in agriculture, which causes environmental pollution (Almaguer et al., 2012). One of the most widely used ways to solve this problem is the use of organic fertilisers that can partially or totally substitute mineral fertilisation. One of the organic sources with the greatest prospects for use is the transformation of solid organic waste by means of earthworms.Humus produced by earthworms possesses a group of physical, chemical and biological properties that make it ideal as an organic fertiliser, as it improves soil structure, increases water retention, facilitates plant nutrient uptake and stimulates plant growth and development, among other functions (Hartwigsen and Evans, 2000; Eyheraguibel et al., 2008). Traditionally, Eisenia foetida S. (Californian Red) and Eudrilus eugeneae K. (African Red) have been the most widely used species in vermiculture and vermicomposting due to their ability to colonise organic waste naturally, their high rate of consumption, digestion and assimilation of organic matter, their ability to tolerate a wide range of environmental conditions and their high reproductive rate. However, studies by Gajalakshmi et al. (2001) and Tripathi and Bhardwaj (2004) show the possibility of using local anecic species in vermicompost production instead of exotic species. For this it is important to create adequate environmental conditions for the earthworms to adapt and express all their potentialities, allowing the production of good quality humus. Among the species with potential is the common earthworm (Lombricus terrestris L.), large and cosmopolitan, which requires optimal conditions for its adaptation to confinement conditions and to obtain acceptable humus yields. Fuentes (1987) suggests the need to build expensive facilities, to avoid that the individuals, due to their natural tendency, abandon the place where they were initially installed, among other reasons, in spite of being present in any Venezuelan yard or farm. Having in into account the stated above see above to following

scientific problem:

The potential of the native earthworm of the Guamita-Pernal Sector, Cojedes State, Bolivarian Republic of Venezuela, for its adaptation to confinement conditions and production of humus with acceptable quality is unknown.

Hypothesis

If the right conditions of humidity, temperature and substrate quality are created for the common earthworm native to the Guamita-Pernal Sector, it will be possible to achieve the adaptation of individuals, their growth and development, as well as obtaining humus of acceptable quality.

General objective

To evaluate the potential of the earthworm native to the La Guamita-Pernal sector for its adaptability to confinement conditions and humus production from different substrates.

Specific objectives:
1. Establish adequate conditions of temperature, humidity, substrate and density of individuals, which allow the native earthworm to adapt to confinement conditions.
2. Evaluate the growth and development of earthworms based on population indicators.
3. Characterise the humus produced from different animal excrements by chemical and physico-mechanical methods.
4. Determine the phytosanitary quality of the humus obtained.

5. Assess the economic feasibility of humus production with native earthworms.

2. LITERATURE REVIEW

2.1. Sustainable agriculture

According to González (2002), Sustainable Agriculture is distinguished by the application of high amounts of organic matter in the support of plant nutrition and in the management of soil fertility conservation.

Organic fertilisation can be an economically and ecologically effective way to reduce dependence on chemical fertilisers, which provide elements that are directly assimilated by plants; however, they can have undesirable side effects, such as eliminating bacteria and micro-organisms that have the function of making the different soil elements assimilable for plant nutrition, contaminate the environment and can affect the health of people and animals due to their excessive use (Mulet del Pozo et al., 2008).

Organic agriculture, on the other hand, has among its most significant advantages that of improving the organic quality and biological activity of the soil, facilitating water penetration, increasing water retention and reducing production costs by reducing the price of fertilisers used. One of the organic sources with the greatest prospects for use is that obtained from the transformation of organic solid waste by means of the earthworm (Almaguer, 2012). With the implementation of organic agriculture, different types of organic fertilisers have been developed, which are obtained according to their origin of formation, by nature or by man's work. The best known types of organic fertilisers are those made up of crop residues, animal manure, ashes, cachaza, natural fertilisers, composts, green manures, liquid fertilisers and earthworm humus (Figueroa, 2002; Téllez, 2007).

2.2. Importance of organic matter

Primavessi (1990) defines organic matter as all dead substances from plants, micro-organisms, animal excretions (from terrestrial fauna, both micro and macro fauna), which are superimposed on the mineral soil in natural environments and which, under the action of edaphoclimatic and biological factors, are subjected to a constant process of transformation. According to Zuñiga (2000), the original source of organic matter in the soil is the remains of plants and animals in different states of decomposition, as well as microbial biomass.Meléndez (2003) considers soil organic matter (SOM) as a continuum of heterogeneous carbon-based compounds, which are formed by the accumulation of partially or completely decomposed materials of animal and plant origin in a constant state of decomposition, of substances synthesised by micro-organisms and/or chemically, of all living and dead micro-organisms and of small animals that have not yet decomposed. It is estimated that the composition of organic matter in the soil environment is defined as 10 % carbohydrates; 10 % nitrogenous compounds including proteins, peptides, amino acids, amino sugars, purines, pyrimidines, 15 % fats, waxes, resins and 65 % humic substances. These percentages are variable and depend greatly on numerous external and internal factors (Schnitzer, 1990).Organic matter is one of the most important factors in maintaining soil productivity in a sustained manner and determines soil fertility. The use of organic fertilisers as an agricultural alternative, arises as a complement and to satisfy the need to restore soils, at least in part, what is extracted from them through agricultural production (Otero, 2010). The importance of organic matter derives from its intervention in important processes such as soil fertilisation, as it improves particle aggregation, water absorption and air content, reduces surface runoff (Bellapart, 1996), facilitates tillage and root development (Honorato, 1993), increases cation exchange capacity and resistance to changes in soil pH (Honorato, 1993; Guerrero, 1996; Bollo, 1999). Organic matter is also important in the development and transformations carried out by soil organisms (Guerrero, 1996) and favours the biological control of pests and diseases. In addition to its effects on the physical, chemical and biological properties of the soil, organic matter is a slow-acting fertiliser material (Bellapart, 1996; Bollo, 1999) and gives rise to biological nutrient cycles in the soil. Moreover, organic matter inputs are critical for long-term soil maintenance and fertility (Chaney et al., 1992; Bohn et al., 1993; Porta et al., 2003; Armenta, 2006). The biostructure and the whole productivity of the soil is based on the presence of decomposing or humified organic matter, which plays an important role in moisture retention, as soil retains only ½ to 1½ parts of water, while organic matter (manure) retains 3/4 parts (Neugebawer, 1992).

2.3. The use of manure as organic fertiliser

2.3.1. Composition of organic fertilisers

A fertiliser is generally considered to be a material that is applied to the soil and stimulates plant growth indirectly by improving the physico-chemical and biological properties of the soil. On the other hand, a material is considered a fertiliser when it stimulates growth in a direct way by supplying essential plant nutrients. Table 1 shows the results of chemical analyses of the main organic fertilisers used in agriculture. It expresses average values, which can serve as a reference for the evaluation of organic fertilisers, but should not be taken as definitive, because they may vary according to their origin.

Table 1. Characteristics of the most frequently used organic fertilisers in agriculture (González et al., 2002).

Analysis	Type of organic fertiliser			
	Manure from sheep	Hen Hen	Manure Bovine	Humus from earthworm
Humidity (%)	61,60	75,00	54,00	56-60
Ratio (C/N)	15:1	22:1	15:03	14:10
M.O. (%)	21,12	15,54	46,1	51,2
N (%)	0,82	0,70	2,18	2,11
P2 O5 (%)	0,21	1,03	0,98	0,81
K2 O (%)	0,84	0,49	1,02	1,53

Design: Álvarez (2013)

The characteristics of organic manures are governed by their organic matter content, the nature of the materials involved in their formation and the fermentation and decomposition process to which the organic waste has been subjected. The chemical composition of manures varies according to the animal's diet. However, nitrogen is one of the most abundant nutrients found in most manures (Miller and Donahue, 1995). According to Van Horn (1995), compost from cattle manure has a similar composition to the source from which it originates, some of the nitrogen is lost during the composting process and minor elements tend to concentrate due to the loss of carbon, oxygen and hydrogen to the atmosphere. However, the organic matter provided by the manure contains appreciable amounts of nitrogen, which can be used by the plant for a long time and the potassium, phosphorus and calcium contained in the manure are considered to be in more or less sufficient and assimilable quantities as a result of microbial activity.

The mineral composition of solid manure is remarkably heterogeneous; it is a compound fertiliser of organo-mineral nature, rich in organic matter with a low content of mineral elements (Labrador, 2001). The nitrogen is almost exclusively in organic form and requires prior mineralisation to be assimilated by the crops. In general, they are characterised by a low content of ammoniacal nitrogen, phosphorus and potassium, which are approximately 50% in organic and mineral form. In addition, they contain large amounts of trace elements and physiologically active substances such as hormones, vitamins and antibiotics, as well as a large population of micro-organisms.

2.3.2. Importance for the development of the plants

It has been shown that the use of organic fertilisers obtained from organic waste from farms or their surroundings, contributes to eliminate environmental pollution produced when they are dumped in the environment, and increases the yields of several species by partially or totally substituting mineral fertilisers (Pérez et al., 1997; González et al., 2002). Organic fertilisers must have a balanced nutrient content of N, P, K, Ca and Mg, so that when used they improve soil fertility and benefit the nutritional status of plants. On the other hand, they should not contain substances that acidify or alkalinise soils and that may affect the normal development of crops. Peña (2002) attributes other functions to organic fertilisers, among them the greater profitability provided by the agricultural system, by greatly reducing irrigation standards, due to their contribution to the greater retention of humidity in the soil, thus reducing production costs. Caballero et al. (2000) studied the effect of organic fertilisers and determined that the use of 10 kg/m² of cow dung and 0.6 kg/m² of earthworm humus, applied separately every two harvests during a rotation of six vegetables, increased yields by over 20 kg/m² and increased the contents of $P O_{25}$, $K_2 O$, as well as the percentage of organic matter in the soil.

2.4. The vermiculture

Tineo (1991), defines vermiculture as: "the breeding and management of earthworms in captivity conditions"; with the basic purpose of obtaining two products of great importance for humans: humus as fertiliser, amendment for agricultural use and protein (fresh meat or flour), as a supplement for animal rations. Therefore, all the different operations related to the breeding and management of earthworms are called vermiculture.

The number of earthworm species described so far is very high; according to Reynolds and Wetzel (2010), there are approximately 8302 species, and on average 68 new species are described each year. For most of these species only the genus to which they belong and their morphological description are known, and their life cycles and ecology are completely unknown. Earthworms represent the largest edaphic animal biomass in most temperate terrestrial ecosystems (Pérez-Losada et al., 2012). These organisms modify soil physical properties such as aggregation, stability and porosity by burrowing galleries (Lavelle and Spain, 2001), and chemical and biological properties such as the rate of organic matter decomposition, nutrient availability, as well as the composition and activity of soil microorganisms and other soil invertebrates (Domínguez et al., 2004, 2010; Lores et al., 2006, Monroy et al., 2008).

Despite the development of vermicomposting, in the last decades there is still insufficient knowledge about the interactions established during the humification process and multidisciplinary studies are required to understand the biological mechanisms that govern vermicomposting, an aerobic process of organic waste degradation and treatment, based on the joint and synergic action of some earthworm species and soil microorganisms (Domínguez, 2004).

There is currently a need to develop appropriate technologies for the production of good quality organic composts, which will enable them to be marketed and used correctly in agriculture. This requires, among other things, methods to evaluate the quality of organic fertilisers, especially those that estimate the concentrations of plant-available elements.

2.4.1. Humus from worm castings

The word humus dates back to 2000 BC and the term has been used since Greek civilisation. For them, humus was that dark brown organic material of a pathogenic consistency resulting from the decomposition of plant and animal remains concentrated in the soil (Theophrastus 372-287 B.C. cited by Fernández et al., 2003). In the humification process oxidation and stabilisation reactions of the organic substrates through the joint decomposing action of earthworms and microorganisms, which convert it into a humified and mineralised material (Domínguez et al., 1997; Bollo, 1999).

Worm castings have a high bacterial flora richness of $2x10^{12}$ colonies per gram of humus produced, compared to the few hundreds of millions present in the same amount of fermented manure. This allows the production of enzymes important for the evolution of organic matter when this material is applied to the soil (Ferruzi, 1986). Worm humus is composed of C, O_2 , N, as well as macro and micro nutrients in different proportions, such as: Ca, K, Fe, Mn and Zn, among others. The final contents per tonne of material will basically depend on the source of origin and the humidity of the material at the end of the process (Fraile and Obando, 1994).

2.4.2. Properties and uses

The use of worm humus is linked to its physical, chemical and biological properties and it is the best known organic fertiliser on the agroecological market. Its composition depends on the substrate on which the worms feed, using organic waste of animal or vegetable origin (Schuldt, 2006). Its use represents an ecological and economic alternative by reducing the amount of urban, agro-industrial and livestock organic waste (Altamirano, 2010). This natural fertiliser corresponds to soil organic matter in a more or less advanced state of stabilisation and consists of humic, fulvic and humin acids. It has physical and chemical characteristics that have positive effects on both the soil and the plant, including: improving soil structure, improving moisture retention, facilitating the absorption of nutrients, and stimulating growth and development by the plant. plant (Mackowiak et al., 2000; Hartwigsen and Evans, 2000; Eyheraguibel et al., 2008). Organic matter is a fundamental part of the vermicomposting process as it is the food for the earthworms. The Mexican Standard NMX-FF-109-SCFI-2007, estimates a value between 20 % and 50 % organic matter on a dry basis (Garg et al., 2006) and reports a higher content in treatments containing cattle manure (29.72%) compared to vegetable waste (23.03%). Delgado et al. (2004) also reported higher amounts of organic matter in treatments containing horse manure (44.9%) relative to sewage sludge (37.5%).

Humus from the decomposition of organic waste with a low C/N ratio and low lignin content is looser and friable compared to compost obtained from plant waste rich in lignin and with a very high C/N ratio. The use of vermicompost is very varied, it can be used as a soil improver or also as a substrate for plant growth, in seed germination, as a support for microbial inoculants, material with the capacity to suppress phytopathogens, to biogenerate degraded soils and even to biorecover contaminated soils in greenhouses or nurseries (Suthar, 2009).

Ortiz and Campos (2002) point out that earthworm humus is an organic fertiliser of extraordinary quality, which has been shown to have a positive effect on the yield and improvement of ferritic soils, with yields of 987 tonnes of earthworm humus.

2.5. The earthworm from

2.5.1. Classification taxonomy

External and internal morphology is used in systematics to classify different species of worms. The parameters used in classification based on external morphology are: number of chetas, number of body segments, position of the clitellum in relation to the prostomium and the characteristics of the prostomium.The species of greatest interest for vermiculture are classified as follows according to Villee (Kooch et al., 2008; Abdullah and Saywack, 2011).

Kingdom: Animalia Phylum: Annelida Class: Clitellata Order: Hepatotoxida Family: Lumbricidae

Genus: Lumbricus, Eisenia

Species: Lumbricus terrestris L, Eisenia foetida S.

2.5.2. Habitat

Worm species of interest for vermiculture are adaptable and can live and reproduce in different ecological conditions. However, like all living organisms, they require well-defined conditions for their optimal development, most of which can be controlled by humans; these are basically: temperature, pH, humidity and adequate feeding. The natural habitat of the earthworm is the soil in which leaf litter, fallen logs and almost any plant material in a state of decay are found. decomposition. Their primary food source is the microorganisms that process this organic matter; their presence, abundance and activity is influenced by the distribution of this substrate in the soil. From an ecological point of view, earthworms are classified into three categories: epigeic, endogeic and anecic, basically according to their feeding strategies and gallery formation (Bouché, 1977; García, 2006). Epigeal: these are earthworms that live on the surface of the soil and feed on leaf litter. They are very mobile and have poorly developed burrowing muscles. Anecic: they feed on leaf litter, which they mix with the soil of the upper horizons. They take refuge in semi-permanent vertical tunnels dug into the soil. They are pigmented on the anterior part of the body and are usually dark brown, large and have strong anterior musculature. Endogeic: they live in the soil and feed on organic matter, as well as live or dead roots, and have well-developed burrowing muscles. According to the use of the resource, they can be classified as: polyhumic, mesohumic and oligohumic.

2.5.3. Characteristics general

The earthworm is an elongated animal with a cylindrical, ringed body and its adult length varies between 5 and 45 cm, depending on the species. Its body is covered with a thin cuticle that protects it from drying out and all its rings (segments or metameres) are the same, except for the first one (prostomium), which contains the mouth, and the last one (pygidium), where the anus is located. Annelids have organ systems, which are included in the body cavity or coelom, containing the coelomic fluid or blood, which acts as a skeleton and is not compressible. At the stage of sexual maturity a zone appears differentiated glandular tissue called the clitellum, which is related t o   reproduction and cocoon laying (Mejías, 2002) (Figure 1).

Figure 1. Anatomy of an earthworm showing its body organs and structures.

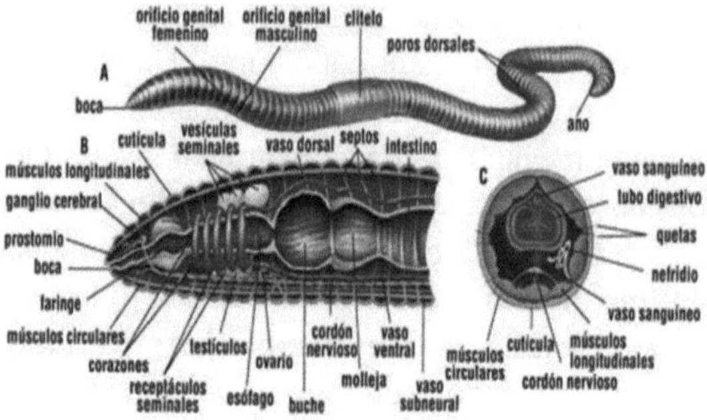

2.5.4. Cycle and conditions of life

The life cycle and conditions vary depending on the species. In general, under favourable conditions each earthworm can produce two cocoons per week, while each cocoon can produce from two to 20 individuals, which after three months develop into sexually mature earthworms. Thereafter they can mate at an interval of seven days and regularly take up to 60 weeks to reproduce. Thus one worm can produce approximately 1500 offspring in one year (Cuevas, 2005).

Eisenia foetida matures sexually at two months of age, which is indicated by the appearance of the clitellum. The mating of two worms takes place with no less than 7 days between each other, from which 1 or 2 cocoons are obtained for each worm. If the medium conditions are optimal, after 14 to 21 days of incubation, the cocoon hatches and 2 to 9 earthworms (usually 2 to 4) are born. feed immediately. Juvenile worms reach sexual maturity between 45 and 90 days after hatching, depending on culture conditions. Eudrilus eugeniae usually produces 1 to 2 cocoons per week, which hatch in 15 to 30 days, with each individual producing 2 to 5 offspring. Similarly, earthworms reach maturity between 32 and 90 days, depending on culture conditions. As mentioned above, earthworm species of interest for vermiculture are adaptable and can live and reproduce in different ecological conditions. However, like all living organisms, they require well-defined conditions for their optimal development, most of which can be controlled by humans; these are mainly temperature, pH, humidity and adequate feeding. When earthworms are in the optimal range of these parameters, they are able to live, reproduce and produce humus.Under unfavourable conditions, worms initially only feed, but do not reproduce, so population growth stops and humus production consequently decreases. If living conditions become more adverse, the worms go into dormancy and will only feed to survive, but neither reproduce nor produce humus. Under these growing conditions, only juvenile earthworms, which are more resistant than adults, will remain. Finally, extreme conditions lead to the imminent death of the worms. Based on the above, a farmer with enough experience in the culture, by visual appreciation of the earthworm population, can detect any problem affecting the development of the culture and can take the necessary measures to solve it in time.

2.5.5. Enemies natural

The main enemy of the earthworm is man. In the wild, the use of herbicides, pesticides and chemical fertilisers cause physiological affectations and disorders that lead to the death of the individual. In farm conditions, the presence of predators is, in most cases, an indicator of incorrect management by the worm farmer, usually low humidity and very acidic beds. Mice, birds and frogs are the main vertebrates that threaten earthworms in breeding sites. They can be controlled by placing nets or protective aerated covers over the beds. In temperate countries, moles are very dangerous and can destroy entire crops. They can be prevented by placing the litter on slabs, strong tarpaulins, metal netting with very narrow mesh, concrete or bricks in order to build a pavement that prevents the mole from getting through.
There are a large number of invertebrates that are predators of earthworms, such as ants, mites, earwigs, centipedes, etc. Ants can cause considerable damage by establishing themselves in high-density colonies in beds with low humidity. In general, these predators are usually controlled by maintaining substrate humidity above 80 % and a pH above 7.0. In tropical and subtropical countries, the planaria (Bipalium kewense Moseley) is undoubtedly the most important pest in flower beds. Infested units may not under any circumstances hand over brood stock to other units until they are fully convinced that they have been eradicated. Hygiene measures for personnel and cleanliness of tools must be emphasised to avoid transferring the infestation to other areas. Increasing the pH of the substrate to 7.5 to 8.0 can control the parasite (Diaz, 2002).

2.5.6. Associated fauna

The decomposition of the organic matter starts from the moment that the deposits are formed with different types of waste for the promotion of vermiculture. Many organisms take part in this process and colonise this substrate to feed on the organic matter or to use it as a shelter. One example are detritivorous organisms such as mealybugs or other arthropods that can compete with earthworms for food, but without causing direct damage. Other invertebrates and micro-organisms that participate in the decomposition of the organic substrate are associated with all crops. It is important to note that in a properly managed crop, none of the mentioned organisms are able to compete or cause significant damage.

2.5.7. Main species used for vermiculture

The most commonly used species are: Eisenia foetida (Californian Red), Eudrilus eugeniae (African Red), Eisenia andrei, Perionyx excavatus and Lumbricus rebellus (Mejías, 2000). Eisenia foetida S. or Californian red worm (Figure 2), has been the most widely used species in vermiculture. In the adult stage, it usually has a length between 5 and 9 cm with a diameter between 3 and 5 mm. It has a characteristic purple or dark red colour and can reach 1 to 1.2 g in weight under optimal conditions. The number of segments varies between 80 and 120 with an average of 95. As adults they have a clitellum or saddle-shaped bulge located between segments 24 to 32.
Eisenia foetida is an extraordinarily prolific, very lively, stress-resistant earthworm (perhaps like no other), and has been found to produce humus at densities of 50 000 to 60 000 earthworms per square metre, which no other wild earthworm can withstand under these conditions. This species lives in captivity without moving from its bed, it matures sexually between the second and third month of life. Every seven to ten days, it deposits a capsule with an average content of 10 eggs and can lay up to 20, which hatch after 14 to 21 days of incubation, giving rise to earthworms that are able to move and feed immediately (Mejías, 2000).

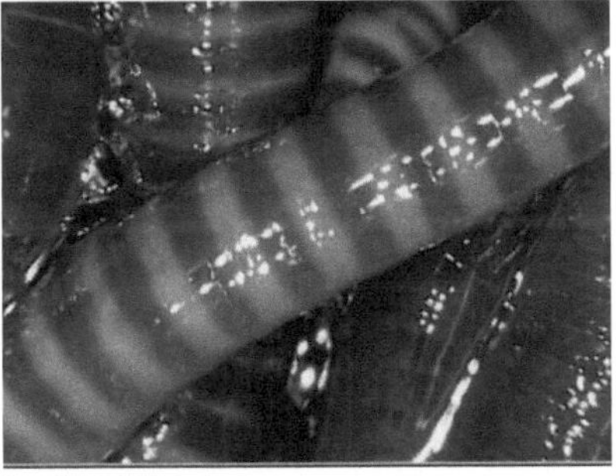

Figure 2. Eisenia foetida. Left: group of adult individuals. Right: worms with coloured banding patterns characteristic of some populations.

Eudrilus eugeniae, or African red worm (Figure 3), is another species widely used in humus production. It is larger than the Californian Redworm and adult individuals reach 15-20 cm. It is dark purplish red in colour and can weigh more than 3 g. Its clitellum is located closer to the prostomium and has a rounded tail (pygidium).

Figure 3. Eudrilus eugeniae. Group of adult individuals.

2.5.8. Importance of earthworms in the decomposition of organic matter

Among the different communities of organisms, the group that has the greatest effect on the modification of the physico-chemical and biological properties of soils is the macrofauna and megafauna, comprising all animals with sizes between 2-20 mm and larger than 20 mm, respectively, including isopods, amphipods, chylopods, diplopods, coleoptera, arachnids and earthworms (Hernández et al., 2012). Earthworms can contribute up to 70% of the invertebrate biomass. In fact, in many tropical savannas earthworm communities play a predominant role and are important regulators of soil structure and organic matter dynamics (Decaëns et al., 2004).
The beneficial effects of earthworms include:

• They improve some physical properties of the soil such as: structure, turbidity, water holding capacity, drainage and aggregate formation and degradation (Blanchart et al., 1999).

• They cause chemical and biological effects on organic matter degradation and nutrient cycling, as well as on the composition and activity of soil micro-organisms and other invertebrates (Lavelle and Spain, 2001).

All these processes contribute significantly to soil fertility by releasing nutrients (P and N) in plant-available forms, which enhances plant growth and crop productivity (Chaoui et al., 2003, Jiménez et al., 2003, Baker, 2007). The natural decomposition of organic waste is a process that takes time, therefore vermicomposting is a mechanism that contributes to accelerate this process, the capacity of the earthworm as an additional organism to these biotechnological systems is of great importance, Since ancient times the earthworm was known as "plough" or "intestine of the earth" (Aristotle) and until the end of the last century, in the first edition of the book "The formation of plant humus", Darwin explains the role played by earthworms in the transformation of the soil (Luévano et al., 2001; Delgado et al., 2004). The raw material used during vermicomposting is organic waste: manure, crop residues, sewage, domestic waste, sludge, etc. These residues are decomposed into humus rich in nutrients (Moreno et al., 2005). The use of manure in vermicomposting is variable, as they are very heterogeneous materials due to the influence of the degree of decomposition of the material, the type of animal they come from, as well as the handling of the animals and manure. Garden or crop residues can be enriched with other materials whose nutrient content is more favourable, e.g. manure and urine (Sanchez et al., 2005). Earthworms prefer soils with abundant clay and with high organic matter content, and are generally scarce in both sandy and strongly acidic soils. The ability of earthworms to bore into the soil to deposittheir excretions inthe galleriesthey form, has a favourable influence on the physical properties of the soil, favouring the formation of stable aggregates. Earthworms ensure aeration and favour water infiltration. It is estimated that 5 % of the total volume of agricultural soils is bored, which allows a significant increase in porosity. During their movement, earthworms mix organic matter with the mineral components of the soil, which creates favourable conditions for plant growth. It can be affirmed that earthworms are real natural farmers, capable of tilling and fertilising the soil. The presence of earthworms and earthworm castings on the soil surface is a symptom of the activity and fertility of the system. Worms also influence the chemical characteristics of the soil, because once they ingest and digest the organic matter, they deposit it in the soil, which allows a better distribution of the nutrients available for the plants. With their movement, earthworms redistribute these elements and other materials from the surface to the depths of the soil, and vice versa. The earthworm population in soils increases considerably as a consequence of conservation management, mainly with the introduction of reduced and shallow tillage systems, as well as with rotation and the use of cover crops. It also increases with increased organic fertilisation and decreased use of agrochemicals. The number of galleries that earthworms are able to open in a soil depends on the species, as well as on the soil characteristics, the type of tillage used and the application of mineral fertilisers and other agrochemicals. Earthworm excreta, which are composed of a complex of organic matter and digested soil, are usually found on the soil surface or in galleries. It is known that earthworm excreta are composed of a high proportion of humus-clay complex, which gives the soil a higher moisture retention and greater protection against erosion. It is generally considered that, of the complex soil fauna, the earthworm is the animal that consumes the greatest amount of organic matter. When there is a high concentration of earthworms in the soil, the processes of mineralisation and humus synthesis are accelerated, due to the action of the micro-organisms associated with the excreta of these animals.

Among the main characteristics of earthworms that favour the development of vermiculture systems are the following:
• They are ubiquitous and colonise various organic wastes naturally.

• They tolerate a wide range of temperatures and humidity.

• They are strong, resistant and easy to handle.

• They have a high reproduction rate.

• They are effective colonisers of all types of environments rich in organic matter.

• They can be adapted a living at captivitywithout escaping from its regardless of climatic conditions and altitude.
• They consume a daily amount of waste almost equivalent to their own weight.

3. MATERIALS AND METHODS

3.1. Materials

The organic material used for the preparation of per compost was cow manure, sow and chicken manure, which were obtained from the waste of the "Maríantonia" farm located in the sector of La Guamita, Cojedes State, where the native earthworms used to obtain the humus were collected.

3.2. Characteristics of the area experimental

The research was carried out in the integral farm Mariantonia, in the sector "El pernal" of La Guamita, Tinaquillo municipality, Cojedes state, in the Bolivarian Republic of Venezuela, with the following geographical coordinates: north latitude 9° 50` 33,9" west longitude 68° 20' 54,9" and an altitude of 401 ml/n.m. The region has typical characteristics of the tropics (warm zone) with a rainy subtropical climate with annual temperatures between 26 and 28°C. Average rainfall was between 1,600 and 1,100 mm with a dry season from November to April and a rainy season from May to October. The highest evaporation rates occur in February, March and April. In the rest of the year they remain uniform. The average annual relative humidity is 62 %. The vegetation present is made up of dry tropical forest species.

3.3. Study of the development of the native earthworm in different substrates

The research was conducted through the following sequence of steps (Figure 4):

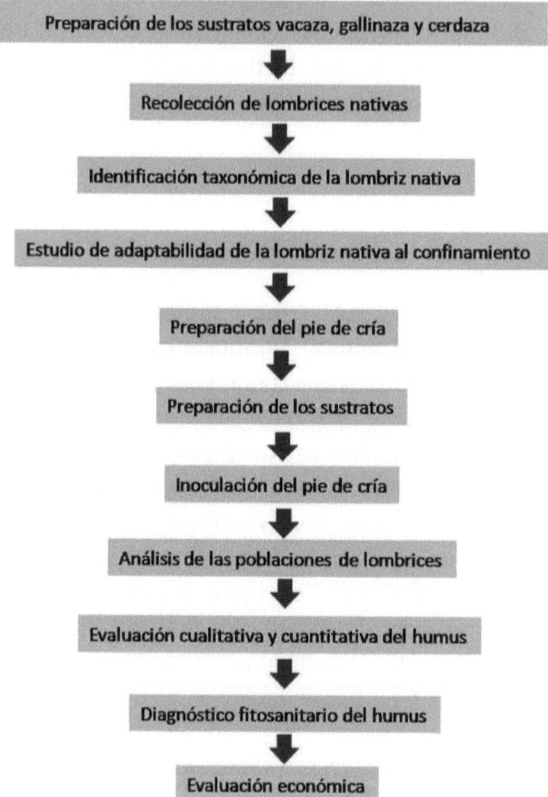

Figure 4. General scheme of work for native earthworm adaptation, humus production and humus quality assessment.

Design: Álvarez (2013)

3.4. Preparation of the substrates vacaza, gallinaza and cerdaza

The cow, hen and sow manures were separated from the collection site and placed in shaded and humid conditions for 20 days, with the objective of stabilising the pH near a range of 6.7 - 7.3 (Pineda, 2006). After this time, the substrates were placed in a container containing earthworms to check if they were introduced into the medium and fed, as an indicator that the substrate was in good condition.

3.5. Collection of native earthworms

For the collection of native earthworms, plastic nets (vegetable sacks) were used with substrate composed of cow manure and chicken manure in a 1:1 ratio. The bags were placed in shaded areas with high humidity. After 30 days, the bags were removed and the worms were extracted and kept in optimal feeding conditions until the beginning of the experiment.

3.6. Taxonomic identification of the native earthworm

Identification of the native earthworm was carried out with the aid of a dichotomous key for the identification of species of the family Lumbricidae (Lainez and Jordana, 1987). Individuals were observed using a stereoscope (Colpostar v6). In addition, a comparative analysis of the observed characters was carried out with those described for the common species used in lumbriculture and present in the country. The analysis was based on the observation of the following morphological and behavioural diagnostic characters (Abdullah and Saywack, 2011):
• Body colour.

• Body length.

• Number of segments.

• Peristomium shape.

• Shape, length and position of the clitoris.

• Habitat (epigean, anecic and endogean) (Garcia, 2006).

• Movement as it moves along the substrate.

3.7. Study of the adaptability of native earthworms to confinement.

In order to establish the appropriate conditions for keeping native earthworms in confinement, an experiment was designed to evaluate adaptability. For this purpose, 10 plastic containers with a capacity of 20 L were prepared with a hole in the base for liquid drainage (Figure 5), in which 0.5 kg of gravel and 0.5 kg of sand were placed, and a uniform layer of clay loam soil 5 cm high was placed on top. On top was placed a 1:1:1 mixture of vacaza-gallinaza-hojarasca.

Figure 5. 20 L plastic containers (0.03 m^3) used for the study of earthworm adaptability in confinement.

Design: Álvarez (2013)

Twenty adult earthworms (density of 1330 individuals m^{-3}) were inoculated in each container and fed every 10 days with a mixture similar to the one placed in the top layer of the container. The substrate was moistened every 2 days to maintain an adequate humidity of approximately 75 % according to the method described by Restrepo (1996) and an average temperature of 26 °C (Giulietti et al., 2008), which was determined with the use of a thermometer (Figure 6).

Figure 6. Measurement of substrate temperature using a thermometer.

Similarly, pH was monitored to keep it in an optimal range between 6.5 and 7.5 (Schuldt, 2001), which was determined by the semi-quantitative litmus paper method. For this, the paper was introduced into the organic matter, pressure was exerted on it to moisten it and then the reading was taken (Zarela et al., 1993). Worms were monitored every 2 days to determine their presence by manually removing the substrate. At 120 days after inoculation, the worms were counted and divided into four categories according to their length (<2.5; 2.5-4.5; 4.5-5.5; >5.5).

3.8. Preparation of the foot of breeding

Once the conditions in which the native earthworms adapted to confinement were determined, the breeding stock that will be used to obtain humus was encouraged. Four plastic containers with a capacity of 100 L (0.15 m^3) each were used (Figure 7), to which layers were added with the same elements of the previous test in the following quantities: 10 cm of gravel (at the base of the container), 10 cm of sand (in the middle) and 10 cm of a mixture of soil-vacaza-ploughsoil in the upper part in a 1:1:1 ratio.

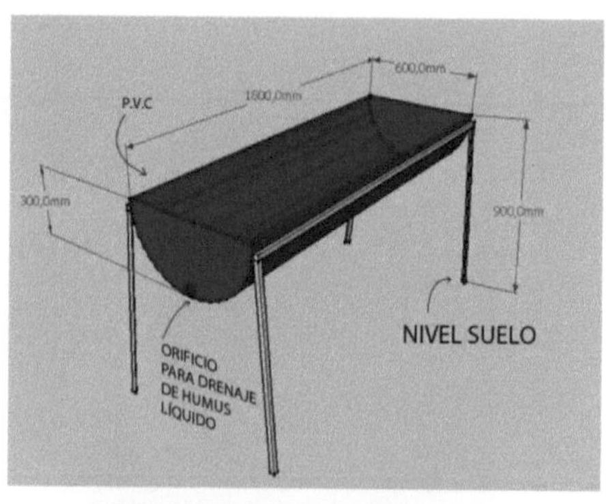

Figure 7. Containers used for brood stock preparation. Left: diagram, right: photograph with substrate.
Design: Álvarez (2013)

Subsequently, 100 worms were placed in each container (1000 individuals m^{-3}) and the containers were placed in a shaded area to lower the temperature and retain the humidity of the substrate. The worms were fed for four months at 10-day intervals with a 5 kg mixture of cowpea - chicken - leaf litter, in a 1:1:1 ratio, and the substrate was watered approximately every three days to maintain the humidity between 70 and 75%.

3.9. Substrates used for the production of humus

Three types of substrates (treatments) were used with the following proportions:
A. Chicken manure: 5 Kg (chicken manure): 1 Kg litter.

B. Vacaza: 5 Kg (vacaza): 1 Kg leaf litter.

C. Bristle: 5 kg (bristle): 1 kg litter.

3.10. Inoculation of the foot of rearing

3.10.1. Packaging

For earthworm reproduction, pyramid-shaped plastic containers (Figure 8, Annex 1) were used (three per treatment) with a capacity of 0.1 m^3, containing 26 kg of dry soil at the base and 6 kg of the excreta/leaf litter mixture in the proportions described in the previous section. After 30 days, 4 kg of a 3:1 excreta/leaf litter mixture was added.

Figure 8. Vessel used for native earthworm reproduction. A: Photograph showing the place where they were kept during the experiment, B: flattened photo of the container.
Design: Álvarez (2013)

3.10.2. Inoculation

An average of 50 individuals of 5 cm long and 3 mm thick were coloured in each container and randomly distributed. The containers were placed during the three months of the study in conditions of shade and optimum humidity, for which sprinkler irrigation was carried out to maintain the humidity of the substrate between 70 and 75 %.

3.10.3. Analysis of the stocks obtained

The worms were extracted from the substrates 90 days after inoculation. A technique similar to the one used for the capture of the initial individuals was applied (section 3.5). The population indicators that were analysed were the following:
• Total individuals: determined from 1 kg of substrate taken at random and multiplied by the total weight of the substrate.

• Reproduction rate: expressed as the ratio between the total number of individuals observed and the number of worms inoculated per treatment.
• Percentage of individuals by stage: juveniles (), sub-adults () and adults ().

3.11. Qualitative and quantitative assessment of humus

Two kg of humus were collected for each treatment. Fractions of 0.66 kg of humus corresponding to each replicate were taken from each treatment and sheep humus was used as a control. The samples were transferred to the Soil Laboratory, Soil and Water Department, EDIAGRO, San Carlos municipality, Cojedes State, for the determination of the following elements:

Mechanical analysis:
• Percentage of sand.

• Percentage of silt.

• Percentage of clay.

Chemical analysis:
• Phosphorus (ppm).

• Potassium (ppm).

• Calcium (ppm).

• Organic matter (%).

• pH.

• C.E.
The determinations were carried out according to La Salle SOP (2013).

3.12. **Phytosanitary diagnosis of humus**

For the phytosanitary analysis, samples were taken from the three treatments, which were transferred to the Laboratory of Phytosanitary and Zoosanitary Diagnosis "Paula Correa Rodríguez", belonging to the National Institute of Integral Agricultural Health, Cojedes State (INSAI, 2013). The presence of parasitic and free-living nematodes as well as insect pests of economic importance was confirmed.

3.13. **Economic analysis**

The economic evaluation of the work included the analysis of the net benefit (total profit - total costs) of the experiment, as well as the humus production in one year, during which 4 productions of one tonne each can be made, as the humification period takes three months. In order to make the calculations, the costs of inputs, labour costs and chemical, mechanical and phytosanitary analysis of the samples were taken into account. Similarly, the profit from the production of the worm humus obtained based on current market prices was considered. Finally, a qualitative assessment of the environmental and social impact of humus production was carried out.

3.14. **Experimental design and statistical processing**

The trials were conducted according to a completely randomised design with three replicates. For the morphological analyses, 20 individuals per treatment were taken. Chemical and mechanical humus analyses were carried out on three different samples per treatment. The data were processed according to the Statgraphic plus 5.1 package on WINDOW, determining the fit to a Normal Distribution by means of the Kolmogorov-Smirnov Goodness of Fit test and the Homogeneity of Variance by means of Bartlett's Tests (Sigarroa, 1985). Specifically, the Analysis of Variance and Tukey's Least Significant Difference (LSD) test and a significance level of a = 0.05 were applied. Where data met the required criteria, they were processed using simple rank ANOVA and Duncan's Multiple Range Test, as appropriate. Data that did not meet these assumptions were compared using the Kruskal-Wallis Test and the Student-Newman-Kwels (SNK) Multiple Range Test.

4. RESULTS AND DISCUSSION

4.1. Collection and classification of native earthworms

From the mixture of cows' milk and goosegrass contained in mesh bags, 1323 individuals in different stages (juveniles and adults) and with lengths ranging from 3 to 10 cm were collected in a period of 30 days. From the characteristics of the clitellum, the native species was identified as Lombricus terrestris L., known as the common earthworm and belonging to the family Lumbricidae (Lainez and Jordana, 1987). Members of the family Lumbricidae are morphologically characterised by the presence of a multilayered clitellum and this structure is of great importance for taxonomic classification of annelids (Pérez-Losada et al., 2012). In order to corroborate the information obtained using the dichotomous key, other aspects of native earthworm morphology and behaviour were analysed. The characteristics observed are shown below (Figure 9):

• Body colour: dark reddish-brown.

• Body shape: sigmoidal.

• Average body length: 15 cm.

• Number of segments: 125

• Shape, length and position of clitellum: 15 - 20 prostomial segments, cylindrical.

• Number and position of setae: arranged in pairs in ventral or ventro-lateral position on each segment.

• Habitat: anecic with developed burrowing apparatus.

• Movement: slow.

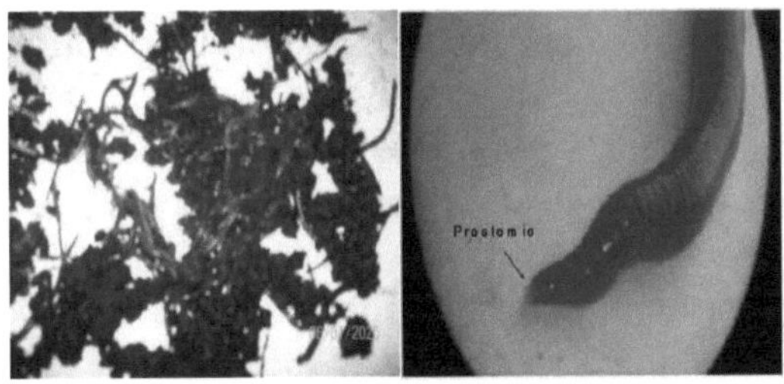

Figure 9. Photograph of the native earthworm collected. Left: population of juvenile and adult earthworms showing the reddish colour typical of the family Lumbricidae. Right: photograph of an adult specimen, magnification (5X).

These data are consistent with the diagnostic characters of the family Lumbricidae. Among the cosmopolitan species abundant in Venezuela are Lumbricus terrestris (common earthworm), Eisenia foetida (Californian red earthworm) and Eudrilus eugeneae (African red earthworm), the latter being cosmopolitan due to their extensive use in earthworm culture. As a result of the comparative analysis, the species Eudrilus eugeneae could be discarded, as it belongs to the family Eudrilidae, with different characters such as the ring-shaped distribution of the mushrooms around the body, which was not observed in the native species (Edwards and Bohlen, 1996). The characters observed also ruled out California red (also belonging to the family Lumbricidae), as it has a characteristic purple, or dark red, colour. In addition, when the body of the animal was pressed, the typical foul odour of California red, which gives the species its name, was not detected (Edwards and Bohlen, 1996). On the other hand, both the Californian red and the African red are epigeal (Ismail, 1997, 2005), whereas the native red showed an anecic behaviour typical of the common earthworm Lombricus terrestres L. (Römbke et al., 2005; Domínguez and Gómez-Brandón, 2010). Additionally, a low reproductive rate typical of this genus was observed. It is important to note that although the family Lumbricidae has been studied for more than 130 years for its ecological importance, there is still no consensus on the classification of the family Lumbricidae, with a group of genera proposed between 6 and 14 in classical studies (Omodeo, 1956; Bouché, 1972), while recent research considers the family to comprise between 31 and 45 genera (Qiu and Bouché, 1998; Csuzdi and Zicsi, 2003).

4.2. Adaptation of native earthworms to confined conditions

The period of preparation or adaptation of the brood stock, which lasted 120 days, showed that the conditions of humidity (75 %), temperature (26°C), feeding and density (1330 individuals m^{-3}) provided ideal conditions for the growth and development of the earthworms. From a total of 200 inoculated individuals, 1657 worms were obtained, distributed as shown in Figure 10.As can be seen, the smallest number of individuals was found in the length range above 5.5 cm, while in the remaining ranges the distribution was homogeneous. An important aspect was that no individuals with lengths greater than 5.5 cm were found in three of the sampled containers, which may be an indication that there was migration or that they did not fully adapt to the confinement conditions.Another significant aspect was that no mortality was observed, which is evidence that the conditions were not stressful and the worms fed adequately.

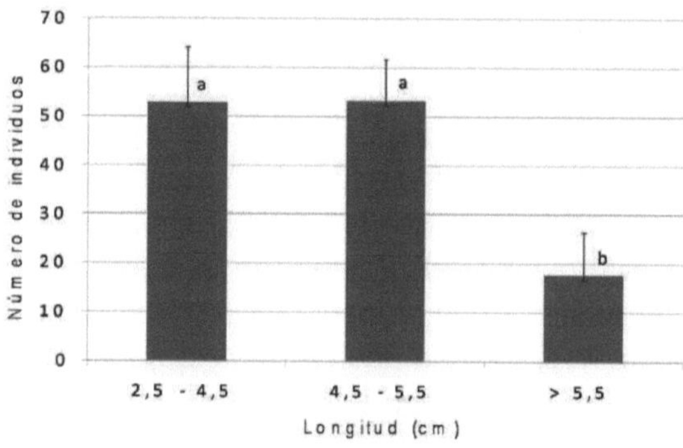

Figure 10. Distribution of individuals by length range after 120 days of rearing in confined conditions.

4.3. Breeding of breeding stock

In order to achieve greater adaptability and prepare the earthworms for humus production, the density of individuals was reduced (1000 individuals m^{-3}), with the same substrate, humidity and temperature conditions (Figure 11).

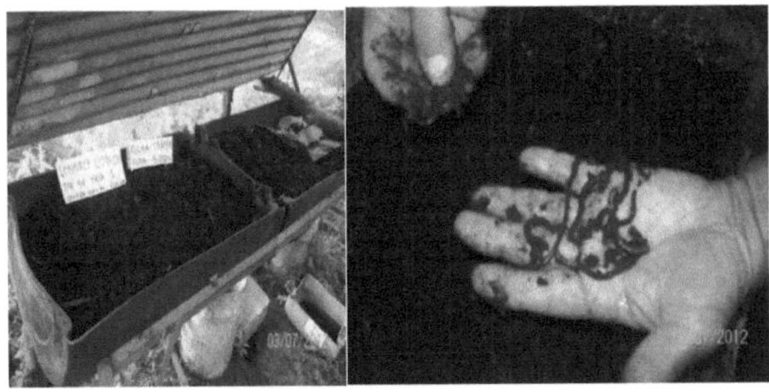

Figure 11. Containers used in the second period of worm adaptation.
Left: containers with substrates, right: adult individuals obtained 120
days after inoculation.

At the end of the experiment, a homogeneous composition of individuals
at different life stages could be observed in the different containers used,
which may be an indication that density and time play important roles in
the adaptation to confinement of these worms. Bernier and Ponge
(1998), in a comparative study on the effect of density on humus
production using Lombricus terrestris, under natural conditions and in
containers, obtained the highest yields using densities in a similar range.
This research showed that by controlling density it is possible to obtain
similar results both in containers and in field conditions.

4.4.   Evaluation of the stocks obtained

Figure 12 shows the total number of individuals 90 days after inoculation.
As can be seen, the best results were obtained in the vacaza treatment
with an average of 3428, followed by poultry manure (1224) and sow
(112). Breeding rates were 34.3:1, 12.3:1 and 1.1:1, respectively. The
superior results obtained with cow dung may be associated with the
digestibility of this material as it comes from a ruminant, in which the
rumen microflora plays an important role. In the case of the sow, a low
earthworm population was obtained, which may be associated with the
presence of planarians that were observed in the substrate. This parasite
grows on acidic substrates and adheres to the worm epithelium sucking
the body fluids causing the death of the animal (Diaz, 2002); Niño and
Robert, 2013). Other predators detected were chylopods, which are also

natural enemies of earthworms. Similar results were obtained by García et al. (1996) and Rodríguez (1999) in a comparative study of humus with different substrates such as vacaza, poultry manure, pig manure, leaf litter, etc., and using the Californian red earthworm for this purpose. In both cases, the highest number of total individuals was obtained in the substrate with vacaza, statistically superior to the rest of the substrates used.

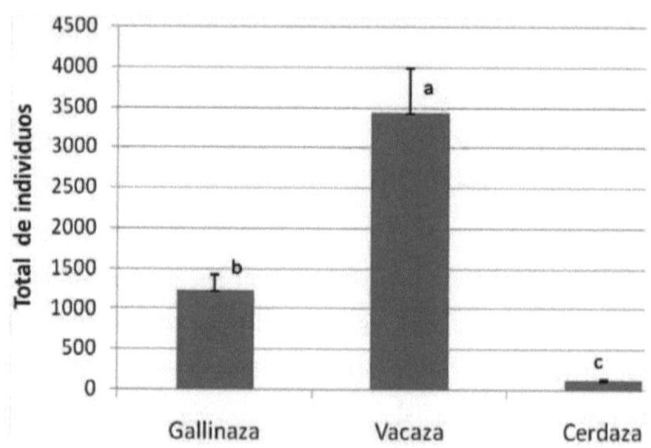

Figure 12 Total of individualsper treatments. Letters different indicate significant differences according to Duncan's test (a<0.05).

The study of the percentage of individuals in the different stages: juveniles, sub-adults and adults is shown in Figure 13. As can be seen, there were no significant differences between stages for the same substrate, nor between the same stages of different substrates or treatments. These results show that the adult individuals inoculated in the different substrates were able to feed and reproduce in such a way that in each substrate an equity between the different life stages of the earthworm was reached (Figure 14). Likewise, numerous cocoons could be observed in both treatments as an indicator of the reproductive potentials reached by the worms in the substrates used.

Gallinaza

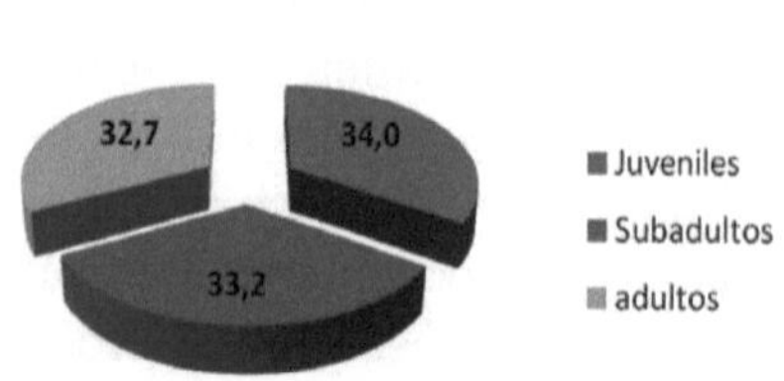

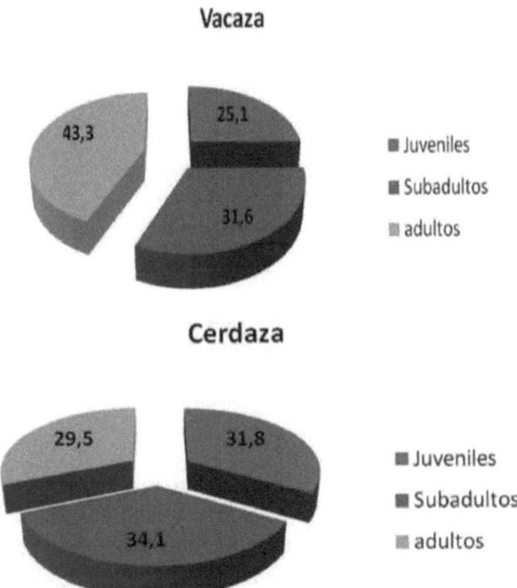

Figure 13. Percentage of individuals in different stages obtained in t h e substrates vacaza, gallinaza and cerdaza.

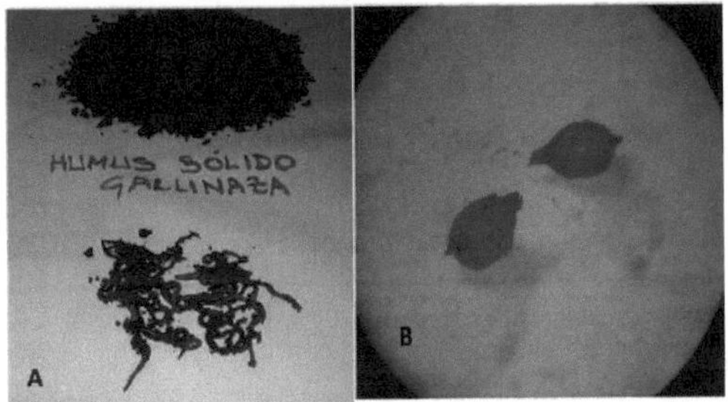

Figure 14. Worms and cocoons observed in the chicken manure substrate at the end of the experiment. Right: juveniles and adults. Left: cocoons.

4.5. Mechanical and chemical evaluation of humus
4.5.1. Mechanical analysis

As a result of the earthworm activity, approximately 24 Kg were obtained (12 Kg from cow manure and 12 Kg from chicken manure) with a percentage of biotransformation of the substrate into humus of 87,5. In the case of the sow substrate, due to the low growth and development of the earthworms, no humus was obtained. The humus obtained from both cow and poultry manure was dark brown, granular and homogeneous (Figure 15).

Figure 15. Earthworm humus from the vacaza and chicken manure treatments, showing the characteristic dark brown colour.

This dark brown colouration observed may be an indication that there was a mixing of organic matter and soil in the earthworm's digestive tract, which may be associated with the behaviour of earthworms, which are considered anecic, i.e. they feed on organic matter present on the surface, but also mix it with the soil, as they have a developed burrowing musculature that allows them to make semi-permanent vertical tunnels. The data obtained from the mechanical evaluation are shown in Table 2. The results show a consumption of sand, silt and clay by the earthworms from the soil placed at the base of the bin, which indicates that suitable conditions were created for the earthworms to have a normal behaviour, penetrating the soil layer and performing the humification function of the substrate.

Table 2. Mechanical analysis of humus from treatments (Annex 2).

Design: Álvarez (2013)

No. Registration	4777	4778	4779
Location	Vacaza	Redfish	Ovejo (Control)
Sand (%) *	76	62	78
Silt (%) ** Silt (%)	16	24	12
Clay (%) ***	8	14	10

USDA system:

*Organic material with particles between 2,00 and 0,05 mm in diameter.

**Organic material with particles between 0,05 and 0,002 mm in diameter.

***Organic material with particles smaller than 0.002 mm in diameter.

These elements play an important role as they are related to soil texture, which is a factor influencing soil fertility and the ability to achieve high agricultural yields.

4.5.2. Chemical analysis

In order to evaluate the quality of the humus obtained, a group of chemical determinations were carried out, such as concentrations of P, K, Ca and organic matter, which play important physiological roles in plants. The results of the chemical evaluation of the humus are shown in Table 3.

Table 3. Chemical evaluation of humus obtained from poultry manure, cow manure and sheep manure as control (Annex 2).
Design: Álvarez (2013)

No. Registration	4777	4778	4779
Location	Vacaza	Hen Hen	Ovejo (Control)
Phosphorus ppm	897	1,344	1,092
Potassium ppm	435	650	500
Calcium ppm	5,077	5,611	6,814
Organic Mat. (%)	23,79	8,9	24,41
pH 1:2.5 (water)	7,8	7,1	8,6
E.C. 1:5 mmhos/cm at 25 °C	0,92	1,19	1,2

Studies on the chemical characterisation of humus using the species Eisenia foetida (Durán and Henríquez, 2007) and Eudrilus eugeneae (Mulet del Pozo et al. 2008), and vacaza as a base material, showed similar results in relation to the organic matter content found in the vacazo humus, but higher than that obtained with humus based on chicken manure.Giulietti et al. (2008) reported lower values of organic matter in cow manure-based humus (13.2 %) and higher values with chicken manure (15.82 %), using in both cases the Californian red earthworm. These results may be related to the fact that cattle feed on grasses, which have a fibre content of between 30 and 35 % (Rodriguez, 1999), compared to manure from other monogastric species. The concentrations of calcium in cow and chicken manure humus reported by Bollo (2001) and Mulet del Pozo et al. (2008) were similar to those evaluated in the treatments. However, the data for phosphorus and potassium concentrations were higher, which could be associated with the chemical composition of the substrates used in these studies. García et al. (1996) and Álvarez-Solís et al. (2010) reported similar phosphorus and potassium values in humus obtained with Eisenia foetida, although organic matter concentrations were 1.96 and 2.1 times higher, respectively. The variations found between different types of humus (quality) may be associated with different factors such as, for example, the concentration of nutrients in the animal and vegetable waste (substrates from which the humus is derived), as it is influenced by the type of diet consumed by the animals, the nature of some vegetable sources that are mixed with animal excreta, the handling and type of storage of the waste used, among others (García et al., 1996; Durán and Henríquez, 2007). These results show that humus based on cow and chicken manure, using Lombricus terrestris L., is also capable of forming humus with acceptable quality, like those obtained with Eisenia foetida, recommended by numerous authors for lumbriculture (Giulietti et al., 2008; Pérez et al., 2012). The chemical components assessed in humus are of vital importance for plant growth and development. Calcium is involved in the regulation of numerous physiological and cellular processes (Hirschi, 2004; Tang et al., 2006). It is an essential nutrient and has functions in metabolic activity such as stabilisation of membranes (Tuna et al., 2007; Taïbi et al., 2012), in the signal transduction as a secondary messenger and in the control of enzyme activity (Laohavisit et al., 2013). Moreover, Ca^{2+} can counteract the adverse effects on plants of abiotic stresses such as soil salinity (Arshi et al., 2006). Phosphorus is the second most important macronutrient essential for plant growth, involved in energy transfer, photosynthesis, transformation of sugars and starch, and movement of nutrients into the plant (Hemalatha et al., 2013). The pH values remained within the recommended range (6.0 - 8.0) (Díaz, 2002; Pineda, 2006) although slightly higher than 7.0, which is the optimum value. These results could be related to the control of humidity, temperature and aeration of the substrates used. When these conditions are adequate, aerobic microorganisms develop, which in addition to participating in the decomposition of organic matter (Nardi, 2002, Domínguez et al., 2010), participate in pH regulation, since when oxygen pressure decreases, anaerobic bacteria proliferate and produce methane, hydrogen sulphide and ammonia as a result of metabolism, which can lower the pH (Díaz, 2002). This aspect is important as bacteria are involved in key processes such as nitrification, sulphur oxidation and nitrogen fixation, which is an indicator for assessing soil health and stability (Perez, 2003). On the other hand, earthworms have Morren's glands, which have the function of secreting calcium carbonate and producing an alkaline digestion, so slightly alkaline pH values are to be expected in the different earthworm humus (Bollo, 1999). The electrical conductivity values remained at low levels when compared to other substrates such as banana and coffee residues and were similar to those obtained by other authors (Durán and Henríquez, 2007), who stated that manure, in general, shows lower K concentrations since most of this element is eliminated through urine. The control of salinity in humus is an important aspect to consider, as values higher than 8 mmhos/cm can cause ionic toxicity to the plant (Mejías, 2000). Excess salts in the substrate can cause metabolic and physiological disorders in plants leading to limitations in processes such as germination (Abbasdokht et al., 2012; Mosavian and Eshraghi-Nejad, 2013), plant growth and development, as well as crop productivity (Saqib et al., 2012; Barakat et al., 2013).

4.6. Phytosanitary evaluation of humus

In order to evaluate the phytosanitary conditions of the humus obtained, the presence of insects and nematodes was determined. The results showed that both the cow manure and chicken manure humus were free of both phytoparasitic nematodes and free nematodes (Annexes 3 and 4). This positive result may be related to the fact that phytoparasitic nematodes can have a free-living phase, during which direct interactions with other organisms such as earthworms can limit their population density (Clermont-Dauphin et al., 2004). Indirectly, earthworms can also influence nematodes, as earthworm activity changes soil structure and resource availability (Blouin et al., 2005). The biogenic structures produced by these ecosystem engineers can alter the physical and chemical properties of the soil, and influence the nematode community by affecting their habitat or food supplementation (Neher, 1999).

In the case of insects, no insect pests were found in the poultry manure humus, while insect pests of no economic importance were detected in the cow manure humus (Annex 3 and 4). These results may be associated with the humidity of the substrate, which was maintained between 70 and 75 %, which allows controlling the presence of pests, ants, etc., which approach the soil because of the sugars produced by the earthworm as it slides through the galleries in the substrate (Pineda, 2006).

4.7. Economic evaluation

Table 4 shows the expenses incurred in carrying out the research. Table 4. Expenditure incurred in the research.
Design: Álvarez (2013)

Inputs and testing analytics	Quantity	Unit price	Total Price (BsF)
Hen Hen	14 Kg	1,20 BsF.Kg-1	16,8
Vacaza	14 Kg	2,00 BsF.Kg-1	28,0
Beer	14 Kg	1,5 BsF.Kg-1	21,0
Plastic tanks (20 L)	10	30 BsF	300
Plastic tanks (100 L)	16	200 BsF	3200
Electricity consumption (Kwatt/h)	180	0.125 BsF	22,5
Chemical-physical analysis of the humus	3	183.3 BsF	550
Total			4138

The amount of humus collected from poultry manure and cow manure was 24 kg. With a market price of 10 BsF.Kg^{-1} , the gross profit calculation (Kg of humus X price per Kg) was 240 BsF, a relatively low value as the production was only designed on a pilot scale. However, there was no net profit (total profit - total expenses) as the costs for inputs and trials were higher than the profit, The net profit calculation made for the production of four tonnes of humus in one year is shown in Tables 5 and 6. As can be seen, the expenses for the production of the first tonne were high due to the initial investment; however, in the remaining productions the expenses were considerably reduced, as only the poultry manure, the cow manure, the electricity consumption for the water supply and the payment for labour were invested. As can be seen, with the production of the first tonne of humus it is not possible to recover the initial investment, due to the current price of inputs on the market.

Table 5. Initial input and labour costs to produce the first tonne of humus. **Design:** Álvarez (2013)

Inputs, labour and production	Quantity	Unit price	Total price (BsF)
Hen Hen	1000 Kg	1,20 BsF.Kg-1	1200
Vacaza	1000 Kg	2,00 BsF.Kg-1	2000
Tank (3000 L)	1	3000 BsF	3000
100 L container	15	200 BsF	3000
Pump ½ HP	1	2800	2800
Hose ½" (20 m)	1	800	800
Shovel	1	350	350
Rake	1	500	500
Forklift	1	2000	2000
Electricity consumption (Kwatt/h)	540	0,125	67,5
Worker	1	800 BsF/month	2400
Total			181175

Price of 1000 Kg of humus = 10000 BsF. Net profit = 10000 - 181175 = - 171151 BsF.

Table 6. Input and labour costs for producing the second tonne of
humus.
Design: Álvarez (2013)

Inputs, labour and production	Quantity	Unit price	Total price (BsF)
Redfish	1000 Kg	1,20 BsF.Kg-1	1200
Vacaza	1000 Kg	2,00 BsF.Kg-1	2000
Electricity consumption (Kwatt/h)	540	0,125	67,5
Worker	1	800 BsF/month	2400
Total			5667,5

Net profit= 10000 - 5667,5 = 4332,5 BsF.

With the second production of one tonne of humus, a profit of 4332.5
BsF is obtained. If we take into account that in the remaining two
productions of the year the same expenses are maintained, we obtain a
total annual profit of 22997.5 BsF, which represents a recovery of the
investment in the year of 13 %. Taking these values into account, the
total recovery of the investment was estimated in a period of
approximately 10 years. An important aspect to highlight is the economic
and environmental benefits of worm humus production, as it is not only a
way to eliminate organic waste produced by chicken excreta and agro-
industrial waste, which represent a source of contamination for animals
and humans, but also has a large group of physical, chemical and
biological properties that improve soil quality and increase crop
productivity (Méndez et al., 2012). This aspect is of great importance
since soil protection is a premise for the sustainable development of
mankind. The indiscriminate use of chemical fertilisers and pesticides,
with the aim of increasing agricultural yields, has contributed to soil
erosion and loss of soil fertility in different regions of the world (Alo et al.,
2013); as well as to the water pollution that can affect human health
through constant exposure to these pollutants (Glasson et al., 1999). It is
known that humic substances contained in earthworm humus can also
stimulate plant growth through hormonal effects (Arancón et al., 2006) or

by increasing plant resistance through improved nutrition (Acevedo and Pire, 2004). In this sense, Tarang et al. (2013) proposed that the combination of biological and chemical fertilisers applied to crops can reduce environmental pollution and the indiscriminate use of chemicals. Moreover, a protective function of humus against pests and diseases affecting crop roots has been described (Yardim et al., 2006). This is important, as the use of vermicompost can reduce the use of pesticides and insecticides with a significant environmental impact. Finally, it is necessary to add the social impact of the practice of vermiculture in the territory, as this activity can have an educational dimension, if this work is made an open space for the continuous learning of old and young generations.

5. CONCLUSIONS

- The conditions of temperature, humidity, feeding, substrate and density of individuals allowed the native earthworms to adapt to confinement conditions, with good growth and development.
- A reproduction rate within the range of the species, a homogeneous distribution of individuals per stage and no mortality was observed in the treatments evaluated.
- The contents of P, K, Ca, organic matter and electrical conductivity of the humus obtained from cow manure and chicken manure showed acceptable values, whose variation depends on the physico-chemical characteristics of the substrates.
- No nematodes or insect pests of economic importance were detected in the humus obtained from cow manure and poultry manure.
- Humus production using the native earthworm was shown to be feasible with a high environmental impact.

6. RECOMMENDATIONS

- Study the microbiota (bacteria and fungi) of the humus, as well as the determination of other chemical elements such as N, C, Fe, Cu, Zn, Mn, etc., which allow a better characterisation of the quality of the humus.
- To carry out studies on the population dynamics of native earthworms in confinement conditions, in order to optimise the resources used to obtain humus.
- Evaluate the humus obtained under production conditions.

- To disseminate the results obtained within the agricultural sector, in order to encourage the production of organic fertilisers as an ecological variant for the biofertilisation of crops.

7. BIBLIOGRAPHY

1. Abbasdokht, H., Ashrafi, E. and Taheri, S. 2012. Effects of different salt levels on germination and seedling growth of sesame (Sesamum indicum L.) cultivars. Tech J Engin & App Sci., 2 (10): 309-313.

2. Abdullah, A.A. and Saywack, P. 2011. Identification and classification of earthworm species in Guyana. International Journal of Zoological Research 7 (1): 93-99.

3. Acevedo, I. and Pire, R. 2004. Effects of vermicompost as a substrate amendment for the growth of milkweed (Carica papaya L.). Interscience, 29(5).

4. Almaguer, J., Reyes, V., Reyes, A. and Villa, O. 2012. Evaluation of the effect of liquid humus obtained by three methods, in pot and field conditions, using maize (Zea may L.) and sugar beet (Betta vulgaris L.) respectively. Local Sustainable Development Journal, 5 (15): 1- 6.

5. Alo, M.N., Egbule, U.C.C., Orji, J.O. and Aneke, C.J. 2013. Microbiological Analysis of Soil from Onu-Ebonyi Contaminated with Inorganic Fertilizer. American Journal of Infectious Diseases and Microbiology, 1 (4): 70-74.

6. Altamirano, F. 2010. "Investigation of the microbial load of compost and humus from humus". University of Jujut, Argentina. Available at: https://sites.google.com/site/steelpulse06/investigandola microbiolog%C3%ADadehumus. Accessed September, 2013.

7. Álvarez-Solís, J.D., Gómez, D.A., León, N.S. and Gutiérrez, F.A.. 2010. Integrated management of organic fertilizers and manures in maize cultivation. Agrociencia 44: 572-586.
8. Arancon, N., Edwards, C., Lee, S. and Byrne, R. 2006. Effects of humic acids from vermicomposts on plant growth. EuropeanJournal of SoilBiology, 42: S65-S69.

9. Armenta V.R., 2006.Transformation of organic matter by the earthworm in soils amended with sewage sludge. Master's Thesis. Faculty of Chemistry. UAEM. Mexico.

10. Arshi, A., Abdin, M.Z. and Iqbal, M. 2006. Sennoside content and yield attributes of Cassia angustifolia Vahl as affected by NaCl and CaCl2. Sci Hortic 111: 84-90.

11. Baker, G. 2007. Differences in nitrogen release from surface and incorporated plant residues by two endogeic species of earthworms (Lumbricidae) in a red brown earth soil in southern Australia. Eur. J. Soil. Biol. 43: 165-170.

12. Barakat, N., Laudadio, V., Cazzato, E. and Tufarelli, V. 2013. Antioxidant Potential and Oxidative Stress Markers in Wheat (Triticum aestivum) Treated with Phytohormones under Salt-Stress Condition Barakat et al. / Int. J. Agric. Biol., Vol. 15, No. 5.

13. Bellapart, C. 1996. New organic agriculture in balance with chemical agriculture. Ediciones Mundi-Prensa. Barcelona. Spain. 298p.

14. Blanchart, E., Albrecht, A., Alegre, J., Duboisset, A., Gilot, C., Pashanasi, B. Lavelle, P. and Brussaard, L. 1999. Effects of earthworms on soil structure and physical properties, p. 149- 172. In P. Lavelle, L. Brussaard & P. Hendrix (eds.). Earthworm management in tropical agroecosystems. Wallingford, United Kingdom.

15. Blouin M., Zuily-Focil Y., Tham-Thi,A-T Laffray D., Reversat G., Pando, A., Tondoh, J. and Lavelle, P. 2005. Belowground organism activities affect plant aboveground phenotype, inducing plant tolerant to parasites. EcologyLetters, 8: 202-208.

16. Bohn, H.L., McNeal, B.L. and O'Connor, G.A. 1993. Soil chemistry. Ed. Limusa. México,D.F.

17. Bollo, E. 1999. Vermiculture: a recycling alternative. Quito. Soboc Grafic. 149 p.

18. Bouché, M.B. 1972. Lombriciens de France. Écologie et Systématique (n hors-série), Institut National de la Recherche Agronomique, Annales de Zoologie-Écologie Animale.

19. Caballero, R.; Gandarilla, J.; Pérez, D. and Rodríguez, D. 2000. Effect of organic fertilisers on intensive orchard management. Centro Agrícola 27(4): 18 -22, October - December.

20. Chaney, D.E., Drinkwater, L.E. and Pettygrove, G.S. 1992. Organic soil amendments and fertilizers. University of California, Division of Agriculture and Natural Resources. Publication 21505. 36 p.

21. Chaoui, H., Zibilske, L. and Ohno, T. 2003. Effects of earthworm casts and compost on soil microbial activity and plant nutrient availability. Soil. Biol. Biochem. 35: 295-302.

22. Clermont - Dauphin, C., Cabidoche, Y.M. and Meynard, J.M. 2004. Effects of AgroforestrySystems, 45: 159-185.

23. Csuzdi, C. and Zicsi, A. 2003. Earthworms of Hungary (Annelida: Oligochaeta; Lumbricidae). Hungarian Natural History Museum, Budapest.

24. Cuevas, G.R., 2005. Development of the earthworm Eisenia fetida Sar. In two locations and effect of vermicomposting on maize in Soconusco, Chiapas. Thesis of Master of Science in Tropical Agriculture. Faculty of Agricultural Sciences. UACh. Mexico.

25. Decaëns, T., Jiménez, J.J., Barros, E., Chauvel, A., Blanchart, E., Fragoso, C. and Lavelle, P. 2004. Soil macrofaunal communities in permanent pastures derived from tropical forest or savanna. Agr. Ecosyst. Environ. 103: 301-312.

26. Delgado, M.M., Porcel, M.A., Miralles, R., Beltrán, E., Beringola, L. and Martín, J.V. 2004. Effect of vermiculture on organic waste decomposition. Rev. Inter. Contam. Amb. 20: 83-86. development of horticultural and horticultural species. Doctoral thesis. Universidad Autónoma Agraria Antonio Narro-UL.

27. Díaz, E. 2002. Vermiculture guide. Vermiculture: a production alternative. Available at: . Accessed September, 2013.

28. Domínguez , J. and Gómez-Brandón, M. 2010. Life cycles of earthworms suitable for vermicomposting. Acta Zoológica Mexicana (n.s.) Special Issue 2: 309-320.

29. Domínguez, J. 2004. State of the art and new perspectives on vermicomposting research. In: Edwards, C.A. (Ed.). Earthworm Ecology, pp. 401-424. CRC Press, Boca Raton FL USA.

30. Domínguez, J., Aira, M. and Gómez-Brandón, M. 2010. Vermicomposting: earthworms enhance the work of microbes. In: Insam, H., Franke-Whittle, I., Goberna, M. (Eds). Microbes at work: from wastes

to resources, pp.93- 114. Springer-Verlag, Berlin Heilderberg, Germany.

31. Dominguez, J., Edwards, E. and Subler, S. 1997. A comparision of vermicomposting and composting. BioCycle 38(4):57-59.

32. Durán, L. and Henríquez, C. 2007. Chemical, physical and microbiological characterisation of vermicomposts produced from five organic substrates. Agronomía Costarricense 31(1): 41-51.

33. Edwards, C.A. and Bohlen, P.J. 1996. Biology and Ecology of earthworm. 3rd Edn., Chapman and Hall, London, ISBN: 0412561603, pp. 426.

34. Eyheraguibel, B., Silvestre, J. and Morard, P. 2008. Effects of humic substances derived from organic waste enhancement on the growth and mineral nutrition of maize. Bioresource Technology, 99: 4206-4212.

35. Fernández, O. and Olate P. 2003. Agronomic evaluation of humic substances derived from earthworm humus. Pontificia Universidad Católica de Chile. 52p.
36. Ferruzi, C. 1986. Manual de lombricultura. Madrid. Spain. Mundi-Prensa. 138 p.

37. Figueroa, V.U. 2002. Sustainable use of manure in irrigated forage systems. Unión Ganadera Magazine. Unión Ganadera Regional de la Laguna. Vol. 38:11-12.
38. Fraile, J. and Obando, R. 1994. Vermiculture: alternative for the rational management of banana wastes. Aqua 3 (4):17-22.

39. Fuentes, J.L. 1987. The breeding of the red earthworm. Ministry of Agriculture, Fisheries and Food. Agricultural Publications, p.4. Available at: Accessed October 2013.

40. Gajalakshmi, S., Ramasamy, E.V. and. Abbasi, S.A. 2001. Potential of two epigeic and two anecic earthworm species in vermicomposting of water hyacinth. Bioresource Technology. 76: 177-181.
41. García, M.D., Domínguez, P.L., Martínez, F. and Covas, M. 1996. Obtaining humus and earthworms (Eisenia foetida) from pig waste and sugar industry waste. Revista Computarizada de Producción Porcina, Vol 3, No. 1.

42. García-Pérez, R.E., 2006. The earthworm as a biotechnology in agriculture. Autonomous University of Chapingo, Mexico. 177 p.

43. Garg, P., Gupta, A. and Satya, S., 2006. Vermicomposting of different types of waste using: A comparative study. Biores. Tecnol. 3: 391-395.

44. Giulietti, A.L., Ruiz, O.M., Pedranzani, H.E. and Terenti, O. 2008. Effect of four vermicomposts on the growth of Digitaria eriantha plants. International Journal of Experimental Botany 77: 137-149.

45. Glasson, J., Thervel, R. and Chadrock, A. 1999. Introduction to environmental impact assessment, 2nd edition Mc-Graw hill, USA. pp. 3-6.

46. González, P. 2002 Influencia de la fertilización orgánica en la producción de forraje y Semilla de cavalia ensiformis / E. Nieto, J. Ramírez, Madelín Cruz (L) . - In DC.   Livestock Ecosystem. No. 1 (1), p. 33.

47. Guerrero, A. 1996. Soil, manures and fertilisation of crops. Ediciones Mundi-Prensa. Bilbao. Bilbao. Spain. 206p.

48. Hartwigsen, H. J. and Evans, M. R. 2000. Humis Acid an Substrate promote Seedling and Root Development. Hort Sciencie. 35(7): 1231-1233.

49. Hemalatha, S., Praveen, V.R., Padmaja, J. and Suresh, K. 2013. An overview on role of phosphorus on growth, yield and quality of sunflower (Helianthus annus L.). International Journal of Applied Biology and Pharmaceutical Technology, 4 (3): 48-55.

50. Hirschi, K.D. 2004. The calcium conundrum. Both versatile nutrient and specific signal. Plant Physiology 136: 2438-2442.

51. Honorato, R. 1993. Manual de edafología. Editorial Universitaria S. A. Ediciones Universidad Católica de Chile. 193p.

52. INSAI, 2013. National Institute of Integrated Agricultural Health. National Network of Phytosanitary Diagnosis and Mycotoxin Screening Laboratories. "Paula Correa Rodríguez Phytosanitary and Zoosanitary Diagnostic Laboratory, Cojedes State, Venezuela.

53. Ismail, S.A. 1997. Vermicology the Biology of earthworms. Oriental Longman Limited, India.

54. Ismail, S.A. 2005. The earthworm Bool. Other India Press, Aprusa, Goa, pp: 101.

55. Jiménez, J.J., Cepeda, A., Decaëns, T., Oberson, A. and Friesen, D. 2003. Phosphorus availability in casts of an anecic savanna earthworm in a Colombia Oxisol. Soil. Biol. Biochem. 35: 715-727.

56. Kooch, Y., Jalilvand, H., Bahmanyar, M.A. and Pormajidian, M.R. 2008. Abundance, biomass and vertical distribution of earthworms in ecosystem unit of hornbeam forst. J. Biol. Sci., 8: 1033-1038.

57. La Salle, 2013. La Salle Natural Sciences Foundation. EDIAGRIO. Dept. of Soils and Water, Cojedes State, Venezuela.

58. Labrador, J. 2001. Ministry of Agriculture, Fisheries and Food. Ediciones Mundi-Prensa. Spain.
59. Lainez, C. and Jornada, R. 1987. Contribution to the knowledge of the oligochaetes (Oligochaeta, Lombricidae) of Navarra. Biology Publications of the University of Navarra. University of Navarra S.A., Spain. ISBN: 84-313-0996-2.

60. Laohavisit, A., Richardsa, S. L., Shabalab, L., Chenc, C., Renato, D.D.R., Colaçoa, S.M., Swarbrecka, E.S., Adeeba, D., Shabalab, S. Shangc, Z. and Daviesa, J, M. 2013. Salinity-induced calcium signaling and root adaptation in Arabidopsis thaliana require the calcium regulatory protein annexin. Plant Physiology Preview. DOI:10.1104/pp.113.

61. Lavelle, P. and Spain, A.V. 2001. Soil Ecology. Kluwer Academic Publishers. Dordrecht, Boston.

62. Lavelle, P. and Spain, A.V. 2001. Soil Ecology. London, United Kingdom.
63. Lores, M., Gómez-Brandón, M., Pérez-Díaz, D. and Domínguez, J. 2006. Using FAME profiles for the characterization of animal wastes and vermicomposts. Soil Biology and Biochemistry. 38: 2993-2996.

64. Mackowiak, C.L.; Grossl, P.R. and Bugbee, B.G. 2001. Beneficial effects of Humic acid on micronutrient availability to wheat. Soil science Society American Journal. 65: 1744-1750.
65. Mejías, P. 2000. Manual de lombricultura. Ed. Agroflor lombricultura. Available at: www.lombriagroflor.cl/. Accessed October 2013.
66. Meléndez, G. 2003. Organic residues and soil organic matter. Organic Fertilisers Workshop. Agronomic Research Centre. University of

Costa Rica.

67. Méndez, O., León, N.S., Gutiérrez, F.A., Rincón, R. and Alvarez, J.D. 2012. Effect of earthworm humus application on growth and grain yield of maize crop. Gayana, Bot. 69 (1): 49-54.

68. Miller, R.W. and Donahue, R.L. 1995. Soils in our environment. 7th ed. Prentice Hall. Englewood Cliffs, NJ.

69. Monroy, F., Aira, M. and Domínguez, J. 2008. Changes in density of nematodes, protozoa and total coliforms after transit through the gut of four epigeic earthworms (Oligochaeta). Applied Soil Ecology. 39: 127-132.

70. Moreno, A., Valdés, M.T. and Zarate, T. 2005. Agricultura técnica, Chile, 65: 26-34.

71. Mosavian, S.N. and Eshraghi-Nejad, M. 2013. Effect of NaCl and CaCl2 stress on germination indicators and seedling growth of canola. Intl. J. Farm. & Alli. Sci., 2 (2): 32-37.

72. Mulet del Pozo, Y.; Díaz, M.E. and Vilches, E.E. 2008. Determination of some physico-mechanical, chemical and biological properties of earthworm humus in conditions of the cowshed of the Guayabal farm, San José de las Lajas, Havana, Cuba. Revista Ciencias Técnicas Agropecuarias, Vol. 17, (1): 27-30.

73. Nardi, S. 2002. "Physiological effect of humus substances on higher plants", Soil Biology and biochemestry (34): 1527-1536.}

74. Nengebamer, A.B. 1992. Ecologically Appropriate Agriculture. German Foundation for International Law. Development Centre for Agriculture and Food. - [s.l]: FRG,184 p.

75. Niño, S. and Robert, D. 2013. Obtaining liquid humus by means of the red earthworm. red worm earthworm. Monographs. Available at: http://www.monografias.com/trabajos71/humus-liquido-lombriz-roja-californiana/humus-liquido-lombriz-roja-californiana2.shtmlAccessed October 2013.

76. NMX-FF-109-SCFI-2007. humus de lombrombrombrriz (lombricomposta) especificaciones y métodos de prueba vermicompost (worm casting) specifications and test methods. Mexico.

77. Omodeo, P. 1956. Contributo alla revisione dei Lumbricidae, Arch. Zool. Ital 41.

78. Ortíz Franco, P. and Álvarez, J.A. 2002. Use of biofertilizers in rainfed oats in the Chihuahua highlands. Available. [online] In: Agricultura Orgánica. Available en:

http://www.smcs.org.mx/pdf/libros/agricultura_org.pdf [Accessed: 18 June 2012].

79. Otero, S. and Teresa, O. 2010. Production and evaluation of vermicompost in anthills, Sierra Nanchititla, Mexico, 53h. Diploma thesis (in option to the degree of Licenciado en Ciencias Ambientales), Universidad Autónoma de México.

80. Peña, E., Carrión, M., Martínez, F., Rodríguez, A and Companioni, N. 2002. Manual for the production of organic fertilisers in urban agriculture. INIFAT. p 35 - 37.

81. Pérez, A., Matías, C., González, Y. and Alonso, O. 1997. Technology for the production of tropical grass and legume seeds. Pastures and Forages (CU) 20: 21.

82. Pérez, G., Monroy, B., Pimienta, E., Posos, P., Aceres, V.A., Toral, J.R. and Carreón, J. 2012. Levels of organic fertilisation using earthworm humus in the cultivation of hibiscus Hibiscus sabdariffa L. ScientiaCucba. 14, (1-2): 47-55.

83. Pérez, N. 2003. Agricultura orgánica: Bases para el manejo ecológico de plagas, Centro de Estudios de Desarrollo Agrario Rural de la Universidad Agraria de La Habana, Asociación Cubana de Técnicos Agrícolas y Forestales, Ciudad de La Habana, Cuba.

84. Pérez-Losada, M., Bloch, R., Breinholt, J.W., Pfenninger, M. and Domínguez, J. 2012. Taxonomic assessment of Lumbricidae (Oligochaeta) earthworm genera using DNA barcodes. European Journal of Soil Biology, 48: 41-47.

85. Pineda, J.A. 2006 Lombricultura / José Arnold Pineda, Instituto Hondureño del Café.1st ed. (Tegucigalpa): (Litografía López). 38 p.: photos ISBN 99926- 37-50-1.

86. Porta, C., López, J. and Roquero, L.C. 2003. Edaphology: for agriculture and the environment. Ed. Mundi-Prensa. Spain. 849 p.

87. Primavessi, A.M. 1990. Beneficios de Materia Orgánica en descomposicao y do Humus. p. 124- 126. -- In: Manejo Ecológico do solo Agricultura en regioe tropicais. -Sao Paulo: Ed. Noble,

88. Qiu, J.P. and Bouché, M.B. 1998. Révision des taxons supraspécifiques de Lumbricoidea. Documents pedozoologiques et integrologiques 3: 179-216.

89. Restrepo, J. 1996. Fermented Organic Fertilisers. Experiencia de Agricultores en Centroamérica y Brasil, CEDECO/OIT (First Edition), San José, Costa Rica. p 52.

90. Rodriguez, A.R. 1999. Production and Quality of Organic Compost by means of the Californian Red Worm (Eisenia foetida) and its Reproductive Capacity. Available at: Accessed: September 2013.

91. Römbke, J., Jänsch, S. and Didden, W. 2005. The use of earthworms in ecological soil classification and assessment concepts. ECT Ökotoxikologie GmbH, Böttgerstrasse 2-14.

92. Sánchez, R., Ordaz, V.M., Benedicto, G.S., Hidalgo, C.I. and Palma, D.J. 2005. Change in the physical properties of a clayey soil due to inputs of vermicompost from cachaza and manure. Interscience 30: 765-779.

93. Saqib, Z.A., Akhtar, J., Ul-Haq, M.A. and Ahmad, I. 2012. Salt induced changes in leaf phenology of wheat plants are regulated by accumulation and distribution pattern of Na+ ion. Pak. J. Agric. Sci., 49: 141-148.

94. Schnitzer, M. 1990. Selected methods forte characterization of soil humic substances. In: McCarthy et al. (Ed.): Humic substances in soil and crop sciences. ASA & SSSA. Madison: 65-89.

95. Schuldt, M. 2001. Vermiculture. Teoría y práctica en el ámbito agropecuario, industrial y domés tico. Imprelyf, La Plata, 136 p.

96. Schuldt, M. 2006. Manual of vermiculture theory and practice. Ed. Mundiprensa. Madrid. 188 pp.

97. Sigarroa, A. 1985. Biometría y diseño experimental. Editorial Pueblo y Educación, 733 p.

98. Suthar, S. 2009. Vermicomposting of vegetable-market solid waste using Eisenia fetida: Impact of bulking material on earthworm growth and decomposition rate. Ecol. Engin. 35: 914-920.

99. Taïbi, Kh., Taïbi, F. and Belkhodja, M. 2012. Effects of external calcium supply on the physiological response of salt stressed bean (Phaseolus vulgaris L.) Genetics and Plant Physiology, Volume 2 (3-4), pp. 177-186.

100. Tang, D., Dean, W.L., Borchman, D. and Paterson, C.A. 2006. The influence of membrane lipid structure on plasma membrane Ca2+-ATPase activity. Cell Calcium 39: 209-216.

101. Tarang, E., Ramroudi, M., Galavi, M., Dahmardeh, M., Mohajeri, F. 2013. Effects of Nitroxin bio-fertilizer with chemical fertilizer on yield and yield components of grain corn (cv. Maxima). International Journal of AgriScience AgriScience Vol. 3(5): 400-405.

102. Téllez, V. 2007. Agroecological fertilisers. Coordinated by DESMI,A.C. [online], Available at: www.abonorgadesmi.com [Accessed: September 2013].

103. Reynold, R. and Wetzel, L. 2010. The role of invertebrate ecosystem enginners. Eur. J. Soil. Biol. 33(4): 159-193.

104. Tineo, B.A.L. 1991. Preliminary study of some reproductive aspects of three earthworm species. Ayacucho, Peru; Universidad Nacional de San Cristobal de Huamanga, Peru, p. 1-20.

105. Tripathi, G. and Bhardwaj, P. 2004. Comparative studies on biomass production, life cycles and composting efficiency of Eisenia fetida (Savigny) and Lampito mauritii (Kinberg). Bioresource Technology. 92: 275-283.

106. Tuna, A.L., Kaya, C., Ashraf, M., Altunlu, H., Yokas, I. and Yagmur, B. 2007. The effects of calcium sulphate on growth, membrane stability and nutrient uptake of tomato plants grown under salt stress, Environ. Exp Bot 59: 173-178.

107. Van Horn, M. 1995. Compost production and utilization, a growers' guide. California Department of Food and Agriculture, University of California. Publication 21514. 17 pp.

108. Yardim, E., Arancon, N., Edwards, C., Oliver, T and Byrne, R. 2006. Suppression of tomato hornworm (Manducaquinquemaculata) and cucumber beetles (Acalymmavittatumand Diabotricaundecimpunctata) populations and damage by vermicomposts. Pedobiologia, 50, 23-29.

109. Zarela, O., Salas, S. and Sánchez, M. 1993. Manual de lombricultura en trópico húmedo. Instituto de Investigaciones de la Amazonia Peruana, Iquitos, Pucallpa. P. 53. Available at: Accessed September 2013.

110. Tarango, Z.R. 2000. La materia orgánica en el suelo [online] In: Abonos orgánicos y plasticultura. Available at: http://www.smcs.org.mx/pdf/libros/abonos_org.pdf. [Accessed: 12 June 2012].

I want morebooks!

Buy your books fast and straightforward online - at one of world's fastest growing online book stores! Environmentally sound due to Print-on-Demand technologies.

Buy your books online at
www.morebooks.shop

Kaufen Sie Ihre Bücher schnell und unkompliziert online – auf einer der am schnellsten wachsenden Buchhandelsplattformen weltweit! Dank Print-On-Demand umwelt- und ressourcenschonend produziert.

Bücher schneller online kaufen
www.morebooks.shop

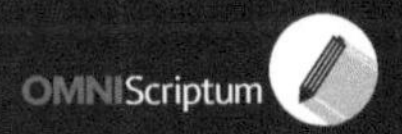

Printed by Books on Demand GmbH, Norderstedt / Germany